THE GRASSLAND BIOME

Colin Grady

Enslow Publishing
101 W. 23rd Street
Suite 240
New York, NY 10011
USA

enslow.com

WORDS TO KNOW

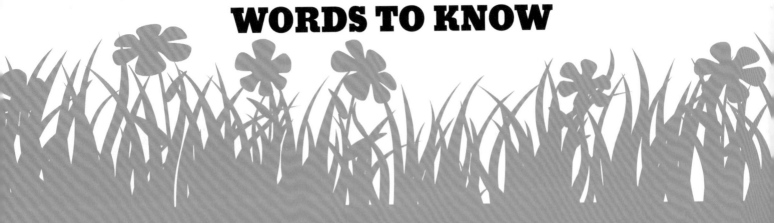

brush fire—A fire that happens in low plants.

climate—The weather conditions in a place over a period of years.

community—A group of living things that share the same area.

continent—One of the seven great masses of land on the earth.

crop—A plant grown on a farm.

equator—A make-believe line around the center of the earth.

graze—To feed on grass.

livestock—Farm animals.

temperate—Neither very hot nor very cold.

termites—Very small antlike animals with soft, pale bodies; they live in large groups.

tropical—Having to do with the hottest and wettest parts of the world.

CONTENTS

1. The Grassland Biome 5
2. A Warm Home: Tropical Grasslands 8
3. Hot and Cold: Temperate Grasslands 13
4. People in the Grasslands 18

Activity: Grassland Animals 22

Learn More 24

Index 24

The grassland biome has lots of wide open land with low plants.

The Grassland Biome

A biome is a community of plants and animals that live together in a certain place and climate. Some biomes include forests, deserts, and the ocean. Another kind of biome is called grassland.

Grassland is land that does not have many trees and is covered mostly with grass. There are two types of grasslands—tropical and temperate. The grassland biome is usually

THE GRASSLAND BIOME

How Many Biomes Are There?

There is not just one answer to this question. Some people say there are just five biomes: water, forest, grassland, desert, and tundra. But most of these biomes can be divided further. For example, there are temperate forests and tropical forests, which are different in many ways. This is why many people disagree about how many biomes there are.

found between the desert and the forest biomes. Grasslands cover more than one quarter of earth's land surface and are found on all continents except Antarctica.

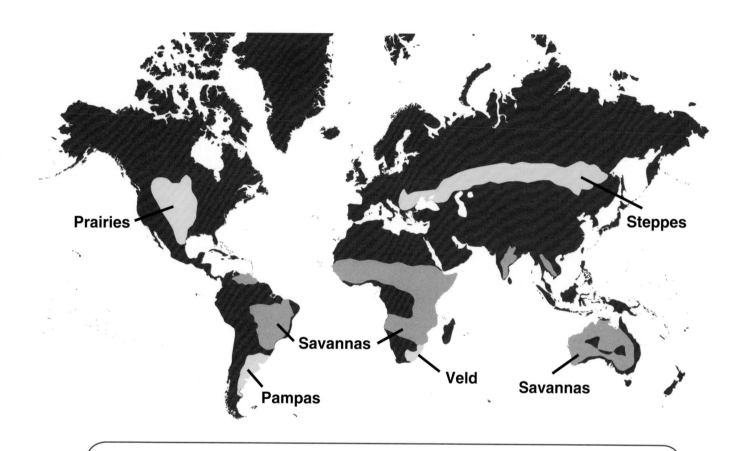

Tropical grasslands, also called savannas, are located around the equator. Temperate grasslands are called prairies in North America, pampas in South America, steppes in Europe and Asia, and veld in South Africa.

A Warm Home: Tropical Grasslands

Tropical grasslands, or savannas, are warm all year long because they are close to the equator. These biomes are found in Africa, Australia, South America, and India.

Animals of the Savanna

Many animals live in savannas all around the world. Some of these animals include elephants, lions, kangaroos, termites, snakes, and beetles. Elephants and lions live in the African

African elephants usually roam the savanna in groups. They eat the grasses, leaves, and fruits that are found there.

savanna. Kangaroos, snakes, and termites are in Australia. The pampas of South America are home to foxes and many rodents. Birds and insects make up most of the wildlife in the grasslands of India.

THE GRASSLAND BIOME

Seasons in the Savanna

Savannas have wet and dry seasons. Some trees and shrubs grow in these grasslands. Savannas get from 10 to 60 inches (25 to 152 centimeters) of rain a year. Dry seasons can last five months or more. During dry seasons, there are many brush fires that destroy the savannas' trees but not the grasses. The grass in a savanna grows back after a brush fire. Savanna grasses have long roots that live through fires. As soon as it rains, the grasses start to regrow.

Plant Lovers

Since savannas are mostly open grass and trees, it is no surprise that many of the animals who live there are plant eaters. Elephants, zebras, and giraffes are just a few of the many plant-eating animals that roam the savannas.

Tiny termites are very common in the Australian savanna—they live mainly on plants. Termite mounds, seen here, have tunnels and rooms where the insects store food.

The prairies of Montana are temperate grasslands that can have very cold winters and hot summers.

Hot and Cold: Temperate Grasslands

Unlike in tropical grasslands, the temperatures in the temperate grasslands can change quite a bit. The summers can be very hot, and the winters can be quite cold. Temperate grasslands are usually flat with very few trees. These biomes are found in parts of North America, South America, Asia, and Africa.

THE GRASSLAND BIOME

Prairies

Prairies are temperate grasslands that have tall grasses and few trees. Prairies usually have rich soil and receive 20 to 35 inches (50 to 89 centimeters) of rain a year. Very few natural prairies remain today. Most prairie land has been turned into farms or used for feeding animals.

Prairie dogs live mainly in the grasslands of North America. They live in large groups and build burrows to protect them from the weather.

Hot and Cold: Temperate Grasslands

Life on the Steppe

Steppes are temperate grasslands found mainly in Europe and Asia that have short grasses and no trees. These are dry areas that receive only 10 to 20 inches (25 to 50 centimeters) of rain a year. Most steppe plants usually grow less than one foot high.

People grow wheat and other crops in this temperate grassland. Also, they let livestock feed on the plants and grasses.

Wide Open Spaces

Steppe means "huge, treeless plain" in the Russian language. It can get very cold and windy in the steppe biome because there are no trees to block the wind. Not many people choose to live in these places.

A farmer herds sheep on the steppes of Mongolia.

Hot and Cold: Temperate Grasslands

One Giant Steppe

The largest temperate grassland in the world stretches all the way from Hungary to China. It goes almost one fifth of the way around the world. It is known by many people simply as The Steppe.

Animals that live on steppes must be able to live without much water. Animals, such as ground squirrels and ferrets, feed on plant life. Other animals, such as bobcats, hawks, and snakes, feed on smaller animals and insects.

People in the Grasslands

The grassland biome is home to many people all over the world. The grasslands' rich soils allow farmers to grow many crops, such as maize, yams, and peanuts in western Africa. Some people who live in grasslands move around from place to place to farm and raise animals. In Mongolia, groups move often to find new places for their horses to graze.

A farmer plows a grassland field in Scotland. Farmers often grow crops in the rich soil of the grasslands.

THE GRASSLAND BIOME

A Biome in Danger

The world's grasslands are in danger of disappearing. Many grasslands have been used up by animals feeding on the grass or from being turned into farmland. When grassland is turned into farmland, there can be problems. Without the roots of the grasses to keep the soil in place, the soil gets blown away. Efforts to save the grassland in the United States include replanting grass where it was cleared for farming. When left alone, the new grass helps turn farmland back into grassland.

Grasslands are very important to humans, as well as all the wildlife that make their home there. We must care for and respect this unique biome.

ACTIVITY
GRASSLAND ANIMALS

You've learned about the animals that live in the grasslands. Now it's time to learn more!

1. Choose any animal that lives in the grasslands. It can be one that you learned about in this book, or you can look for more on the Internet.

2. What do you think life is like for your animal in the grasslands? Do some research online. Write down the answers to the following questions: What does it eat? Who are its enemies? How does it stay safe? How is it suited to the climate? Does it live in big groups or alone?

3. Put together a short picture book that shows how your animal lives in the grasslands. You can draw pictures or print some off the Internet.

4. Include a cover that shows your animal in the biome. Give your book a title—for example, "A Lion's Life in the Grasslands."

5. On each page, illustrate one part of the animal's life in the biome. One page would show the foods it eats. The next would show its enemies, and so on.

6. Staple your pages together and share what you have learned with your class.

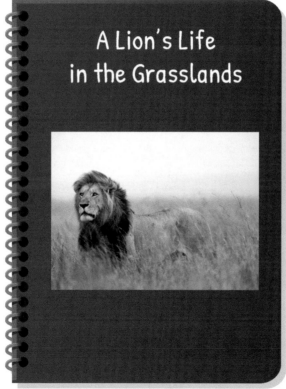

A Lion's Life in the Grasslands

LEARN MORE

Books

Duke, Shirley. *Seasons of the Grassland Biome*. Vero Beach, FL: Rourke, 2014.

Johansson, Philip. *The Grasslands: Discover This Wide Open Biome*. New York: Enslow, 2015.

Patkau, Karen. *Who Needs a Prairie?* Toronto: Tundra Books, 2014.

Royston, Angela. *Grassland Food Chains*. Chicago: Heinemann, 2014.

Websites

Kids Do Ecology
kids.nceas.ucsb.edu/biomes/grassland.html
Photos, facts, and links to information about grasslands.

Ducksters
ducksters.com/science/ecosystems/grasslands_biome.php
Read more about the grassland biome.

INDEX

biome, 5, 6, 8, 13, 15, 18, 20
brush fire, 10
continent, 6
crop, 15, 18
farmland, 20
livestock, 15
prairies, 14
savanna, 8–10
steppe, 15, 17
temperate grasslands, 5, 13–15, 17
tropical grasslands, 5, 8, 13

Published in 2017 by Enslow Publishing, LLC.
101 W. 23rd Street, Suite 240, New York, NY 10011

Copyright © 2017 by Enslow Publishing, LLC

All rights reserved.

No part of this book may be reproduced by any means without the written permission of the publisher.

Library of Congress Cataloging-in-Publication Data
Names: Grady, Colin, author.
Title: The grassland biome / Colin Grady.
Description: New York, NY : Enslow Publishing, 2017. | "2017 | Series: Zoom in on biomes | Audience: Ages 7+ | Audience: Grades K to 3. | Includes bibliographical references and index.
Identifiers: LCCN 2015048570 | ISBN 9780766077805 (library bound) | ISBN 9780766077782 (pbk.) | ISBN 9780766077799 (6-pack)
Subjects: LCSH: Grassland ecology--Juvenile literature. | Grassland animals--Juvenile literature.
Classification: LCC QH541.5.P7 S86 2017 | DDC 577.4--dc23
LC record available at http://lccn.loc.gov/2015048570

Printed in Malaysia

To Our Readers: We have done our best to make sure all website addresses in this book were active and appropriate when we went to press. However, the author and the publisher have no control over and assume no liability for the material available on those websites or on any websites they may link to. Any comments or suggestions can be sent by e-mail to customerservice@enslow.com.

Photo Credits: Cover, p. 1 Chris Wiktor/Shutterstock.com; throughout book, excape25/DigitalVision Vectors/Getty Images (flowers background), Alonzo Design/DigitalVision Vectors/Getty Images (lion and elephant graphics), Alex Belomlinksy/DigitalVision Vectors/Getty Images (grass graphic); p. 4 Mike Hill/Alamy; p. 7 Terpsichores/Wikimedia Commons/Creative Commons Attribution-Share Alike 3.0/Biome map 07.svg; p. 9 Viliers Styen/Shutterstock.com; p. 11 iStock.com/aleskramer; p. 12 iStock.com/jodiecoston; p. 14 Henk Bentlage/Shutterstock.com; p. 16 Dmitry Chulov/Shutterstock.com; p. 19 Pearl Bucknall/Alamy; p. 21 nicolemoraira/Shutterstock.com; p. 23 Maggy Meyer/Shutterstock.com.